Bibliographic information published by the German National Library:

The German National Library lists this publication in the National Bibliography; detailed bibliographic data are available on the Internet at http://dnb.dnb.de .

Imprint:

Copyright © 2017 GRIN Verlag, Open Publishing GmbH
Print and binding: Books on Demand GmbH, Norderstedt Germany
ISBN: 9783668474376

This book at GRIN:

http://www.grin.com/en/e-book/365257/a-physiochemical-analysis-of-two-varieties-of-the-tiger-nut

Adekunle Jelili Olaoye

A physiochemical analysis of two varieties of the Tiger Nut

Characteristics of the Ayaya

GRIN Publishing

GRIN - Your knowledge has value

Since its foundation in 1998, GRIN has specialized in publishing academic texts by students, college teachers and other academics as e-book and printed book. The website www.grin.com is an ideal platform for presenting term papers, final papers, scientific essays, dissertations and specialist books.

Visit us on the internet:

http://www.grin.com/

http://www.facebook.com/grincom

http://www.twitter.com/grin_com

Physicochemical and proximate analysis of two varieties of tiger nut brought to Sabo market

Osogbo, Osun State, Nigeria.

Olaoye , A. J

Department of Chemistry Ladoke Akintola University of Technology, Ogbomoso, Oyo State, Nigeria

Abstract

This study was based on the physico- chemical and proximate analysis of two varieties of tiger nut (brown and yellow). Tiger nut were obtained from Sabo market Osogbo, Osun state Nigeria. The result obtained showed that the specific gravity of the sample were 0.95 and 0.93 for yellow and brown tiger nut respectively, the refractive index of the sample were 1.510 and 1.488, the pH of the sample were 8.85 and 8.06 while viscosity of the sample were 97226.86 and 89377.14 for yellow and brown tiger nut respectively. The chemical properties of the sample showed that, the iodine value of the sample were 11.62 and 15.23, the acid value of the sample were 14.63 and 8.82, peroxide value of the sample were 3.4 and 2.2 while saponification value were 7.91 and 9.82 for yellow and brown tiger nut respectively. The result of proximate analysis showed that the percentage of moisture content were 4.80 and 4.00, the percentage crude protein of the sample were 8.07 and 7.30, percentage of crude fibre of the sample were 7.60 and 6.80, the percentage crude fat were 26.10 and 25.50, the percentage total ash of the sample were 2.03 and 2.00 while percentage carbohydrate of the sample were 51.40 and 54.40 for brown and yellow sample of tiger nut respectively. The tiger nut could be source of industrial materials.

TABLE OF CONTENTS

INTRODUCTION

Tiger nut is commonly known as earth almoud Chufa nut grass and Zulu nut. It is known in Nigeria as Ayaya in Hausa, Ofio or Imumu in Yoruba and Aki - hausa in Igbo where three varieties black, brown and yellow are cultivated (Adejuyitan, 2011). Among these only two varieties yellow and brown are readily available in the market. The yellow variety is preferred to all other varieties because of its attractive colour and fleshier body (Belewu and Belewu 2007). The yellow variety also yield more milk upon extraction, contains fat and more protein and processes less anti-nutritional factors especially poly phenols (Okafor *et al* 2003).Tiger nut is an underutilized sedge of the Family Cyperaceae, Kingdom Plantae, Order Poales, Genus Cyperus, which produces rhizomes from the base and tubers that are somewhat spherical. Tiger nut with the binomial name cyperusesculentus is of several varieties. Tiger nut plant (*Cyperusesculentus L.*) which is of the family Cyperaceae (Adejuyitan, 2011, Belewu and Belewu, 2007). This plant is cultivated for its small tuberous rhizome which is eaten raw or roasted, pressed for its juice to make beverage or milk, extracted of non-drying oil or used as hog feed (Belewu and Belewu 2007). It is variously known in Nigeria as *ofio* or *aki-hausa* in Igbo, *imumu* in Yoruba and *aya* in Hausa (Adejuyitan, 2011, Anon. 2012). Its health benefits were enumerated by (Adejuyitan, 2011).Tiger nut milk originated from Spain where it is known as *chufa dehorchata* while it is commonly called *kunnu aya* in Northern Nigeria (Bamishaiye and Bamishaiye,2011). It is a healthy and rich source of nutrients such as carbohydrates, vegetable fat, protein, fibre, vitamins, minerals, energy and some digestive enzymes such as catalase, lipase and amylase (Adejuyitan, 2011, Belewu and Belewu, 2007, Anon. 2012, Bamishaiye and Bamishaiye, 2011, Arnau, 2009). Tiger nut milk has been reported to contain moreiron, magnesium and carbohydrate than cow milk (Chevallier ,1996). In addition, it has the advantage of not containing sodium, lactose sugar,

Case in protein, gluten, cholesterol and therefore ideal for people who are hypertensive or do not tolerate gluten or lactose and its derivatives present in cow milk (Belewu and Abodunrin, 2006, Anon. 2005). Unlike soymilk or other soy products, tiger nut milk does not produce any allergy (Anon, 2005).

The physico-chemical, nutritional and sensory comparison of animal milk with other plant milks such as soymilk have been copiously reported (Hajirostamloo, 2009., Anon. 2012) but there is a paucity of such information with tiger nut milk. The physical characteristic of tiger nut includes;

4

- It is an annual or perennial plant, growing to 90cm tall, with solitary stem growing from a tuber.

- The flowers are hermaphrodite have both male and female organ and are pollinated by wind.

- The plant foliage is very tough and fibrous

- The plant prefers light (sandy), medium (loamy), and heavy (clay) soils and can grow in heavy clay soil.

The plant prefers acid, neutral and basic (alkaline) soils

- It cannot grow in the shade; it requires moist or wet soil (Facciola,1990).

USES OF TIGER NUT

EDIBLE USES

The tubers are edible, with a slightly sweet, nobly flavour. Tuber – raw, cooked or dried and ground into a powder is also used in confectionery. They taste best when dried. They can be cooked in barely water to give them a sweet flavour and then be used as dessert nut. They are quite hard and are generally soaked in water before they can be eaten, thus making them much soften and giving them a better texture. A refreshing beverage called tiger nut milk or Horchata de chufas is made by mixing the grounded tubers with water, cinnamon, sugar, vanilla and ice (Cantalejo 1997). The ground up tuber can also be made into plant milk with water wheat and sugar. Horchata de chufas can be useful in somehow replacing milk in the diet of people intolerant to lactose. The roasted tubers are a coffee substitute and the base of the plant can be used in salads (Hedrick, 1972)

MEDICINAL USES

Tiger nuts are regarded as a digestive tonic, having a heating and drying effect on digestive system and alleviating flatulence. They also promote urine production and menstruation.

The tubers are said to be aptirodisiac, carminative, diuretic, emmenagogue, stimulant and tonic (Chopra *et al* 1986). In Ayurvedic medicine, they are used in the treatment of flatulent, in digestion, colic, diarrheoa, dysentry, debility and excessive thirst (Chevellier, 1996)

USED AS OIL

Edible oil is obtained from the tuber it is considered to be superior oil that compares favorably with olive oil. Tiger nuts have excellent nutritional quality with a fat composition similar to olives and a rich mineral content, especially phosphorous and potassium. Tiger nut are also gluten and cholesterol free, and have a very low sodium content the oil of the tuber was found to contain 18% saturated palmitic acid and stearic acid and 82% unsaturated oleic acid and linoleic acid fatty acid (Daniel *et al* 2000). The tuber contain up to 30% of a non drying oil, it is used in cooking and in making soap (Rosengartenjnr, 1984). It does not solidify at 0^{OC} and stored well without going rancid (Komaro, 1968). Since the tuber contains oil, cyperusesculentus has been suggested as potential oil crop for the production of bio diesel.

COMMON FATTY ACIDS

FATTY ACID	NUMBER OF CARBON ATOMS	STRUCTURE
Myristic acid	14	$CH_3(CH_2)_{12}$ COOH
Palmitic acid	16	$CH_3(CH_2)_{14}$ COOH
Stearic acid	18	$CH_3(CH_2)_{16}$ COOH
Oleic acid	18	$CH_3(CH_2)_7CH=CH(CH_2)_7COOH$
Linoleic acid	18	$CH_3(CH_2)_4CH=CHCH_2CH=CH(CH_2)_7COOH$
Linolenic acid	18	$CH_3CH_2CH=CHCH_2CH=CHCH_2CH=CH(CH_2)_7$ COOH

SAMPLE COLLECTION AND PRE-TREATMENT

Sample of fresh tiger nut rhizome (yellow and brown varieties were obtain from Sabo Market Osogbo, Osun State, South Western part of Nigeria. The Nuts were dried under sunlight and grounded with the used of mortar and pestle to obtained the flour.

6

APPARATUS

Soxhlet extractor, Water bath, Heating mantle, weighing machine, Dessicator , Mortar and Pestle

Beaker,Burette, Conical flask, Quick-fit Volumetric flasks and Measuring cylinder

CHEMICAL AND REAGENTS

Petroleum Ether (60 – 80^{0}c), Anti bumping granules, Potassium Iodide, Acetic acid, Chloroform, Sodium Thiosulphate, Distilled water, Starch indicator, Sulphuric acid, Phenolphthalein, Iodine Trichloride and Ethanol

PHYSICAL PROPERTIES

REFRACTIVE INDEX

Refractive index of the oil was done using abbe refractometer the surface of the prisms was cleaned up with either. 2 drops of the oil was applied at the lower prism and the prisms were closed up. Water was passed through the jackets at 45^{0}C. The jacket was adjusted for reading to be taken. (Morris, 1999)

DETERMINATION OF SPECIFIC GRAVITY OF THE SOIL

Specific gravity of oil was determined at room temperature. 10cm^3 of the oil sample was weighed as W$_2$. 10cm^3 of distilled water was also weighed and the weight was recorded as W$_1$

Specific gravity = $\dfrac{W_2 + 0.00064\ t^1}{W_1}$

W$_2$ = 0.4266

W$_1$ = 0.4448

t = temperature at which oil is weighed (45^0 C)

$t^1 = t - 15.5^0C$ (Morris, 1999)

Other physical properties

Colour

The colour of the sample was determined by sighting and matching with standard colour.

Viscosity

The viscosity was determined with aid of viscometer tube at room temperature. The sample was put in the tube and rate of flow was recorded.

pH

The MetroHm 632 pH meter was used in determining the pH of the samples. The electrode was firstly placed in a known buffer solution before the pH was determined.

CHEMICAL PROPERTIES

SAPONIFICATION VALUE

2g of oil was weighed into a conical flask, $25cm^3$ of 0.5M alcoholic KOH was added. A blank was prepared by putting $25cm^3$ of the alcoholic KOH in a similar flask reflux condenser was fitted to the flask containing the mixture, this was heated in a water bath for one hour, swirling the flask from time to time. The flask was the allowed to cool a little and the condenser was washed down with a little distilled water, the excess KOH was titrated with 0.5M HCL using phenolphthalein indicator.

The saponification value was calculated from the difference between the blank and the sample titration.

Saponification value = $\dfrac{(a-b) \times f \times 28.05 \times 100}{\text{Weight of sample}}$

Where: a = titre value of sample

b = titre value of blank

f = factor of 0.5M HCL

28.05 = mg of KOH equivalent to 1cm^3 of 0.5M HCL

Weight of sample = 2g (Morris, 1999)

ACID VALUE

0.1g of the oil was dissolved in 2.5cm^3 of 1: 1v/v ethanol: diethyl ether solvent and titrated with 0.1N sodium hydroxide while swirling using phenolphthalein as indicator.

The acid value was calculated using the formula:

Acid value = $\underline{5.61 \times N \times V}$

W

Where: N = the normality of sodium hydroxide

V = the volume of sodium hydroxide in cm^3

W = weight of the sample = 0.1g (Morris, 1999)

ESTER VALUE

This was obtained by finding the difference between the saponification value (S.V) and Acid value (A.V) (Morris, 1999)

PEROXIDE VALUE

5g of the oil sample was dissolved in 30cm^3 of a slovent mixture consisting of 60% glacial acetic acid and 40% of chloroform. 0.5cm^3 of a saturated solution of potassium iodide was added. The flask was shaken until the solution became clear. After 2 minutes from the time of addition of KI, 30cm^3 of distilled water was added and titrated with 0.01N sodium thiosulphate solutions. It was then shaken vigorously to remove the last traces of iodine from the chloroform layer.

Peroxide value (milliequivalent/1000g) = $\dfrac{ML \times N \times 1000}{W}$

(millimole/1000g) $= \dfrac{0.5 \times ML \times N \times 1000}{W}$

Where: ML = titre value

N = normality of $Na_2S_2O_3$ solution

W = weight of the oil sample = 5g (Morris,1999)

IODINE VALUE

1g of oil was weighed into a conical flask. $10cm^3$ carbon tetrachloride was added, then $20cm^3$ of wijs solution was added and the flask was covered, mixed and allowed to stand in the dark for thirty minutes. 10% potassium iodine was prepared by weighing 10g and dissolved it in $100cm^3$ of distilled water. Then $15cm^3$ of the prepared 10% KI solution and $100cm^3$ distilled water were added to the content in the flask. It was mixed thoroughly and titrated against 1.0 N thiosulphate solution. Starch indicator was used and blank determination was carried out under the same condition.

IODINE VALUE (I.V) = $\dfrac{12.69 \times N \times (V_2 - V_1)}{W}$

Where: Weight of oil sample (W) = 1g

V_1 = Volume of thiosulphate used in blank

V_2 = Volume of thiosulphate used in test

N = Normality of thiosulphate solution (Morris, 1999)

Digestion of Sample

About 2g of the pre-treated sample was weighed and heated in a Muffle furnace for 8 hours at a temperature of 550°C, then, 1g weighed out ashed sample was digested with 20 ml of perchloric acid, nitric acid, sulphuric acid and hydrochloric acid respectively. The digested samples were stored in a 100ml volumetric flask prior to analysis. The same procedure was used for the other tiger nut species.

Determination of Percentage Moisture

Moisture content was determined by the method of AOAC (1990). The moisture content was calculated as the loss in weight of the dried sample in an oven until a constant weight was obtained and was cooled in a desiccator.

The percentage (%) moisture content by weight was calculated using formula;

$$\% \ Moisture Content = \frac{W_2 - W_3}{W_2 - W_1} \times 100\% \ ....(i)$$

Where;

W_1 = weight of Petri dish,

W_2 = weight of Petri dish + sample before drying,

W_3 = weight of Petri dish + sample after drying.

Determination of Percentage Crude Protein

The percentage crude proteins of the samples were determined by micro Kjedahl using the method of AOAC (1990). The total nitrogen estimated by the Kjedahl method in a volumetric method is given by the formula below;

$$\% \ Nitrogen = \frac{(Titter Value blank) \times 0.0014 \times vol.of digested sample}{Aliquot aken \times weight of substances of dry matter} \times \frac{100}{1} \ ....(ii)$$

Therefore, $\% \ Crude Protein = \% \ Nitrogen \times 6.25 \ ....(iii)$

Fatty Material Content Determination

Two (2) grams of the sample was placed on a filter paper and put into the soxhlet extractor and extracted into a pre-weighed round bottom flask with low boiling petroleum ether (40 – 600C) using a soxhlet extractor for 8 hours. The solvent was recovered by rotary evaporation, and drying was completed in a freeze dryer. Finally the flask and its content were heated at 900c in an oven for 2 hours, and cooled in a dissector and weighed. The process of heating and cooling was repeated until a constant weight was obtained. AOAC (1990)

$$\% \text{ Fatty materials} = \frac{\text{Weight of fatty material} \times 100 \%}{2 \text{ g}}$$

Ash Content Determination

This was done according to AOAC (1990), where 2g of the sample was weighed (W_1) into pre-weighed empty crucibles (W_0) and place into a muffled furnace at $600°C$ for 3 hours. The ash was cooled in a desiccator and weighed (W_2). The weight of the ash was determined by the difference between the sample, pre-weighed crucible and the ash in the crucible.

Percentage ash was calculated as:

$$\% \text{ Ash} = \frac{W_2 - W_1}{2g} \times 100 \%$$

Crude Fibre Content Determination

Percentage crude fibre was determined by the method described in AOAC (1990) in which 2g of ground sample was weighed (W_0) into a $1dm^3$ conical flask. Water ($100cm^3$) and $20cm^3$ of 20% H_2SO_4 were added and boiled gently for 30 minutes. The content was filtered through Whatman No.1 filter paper. The residue was scrapped back into the flask with a spatula and $100cm^3$ of water and $20cm^3$ of 10% NaOH were added and allowed to boil gently for 30 minutes. The content was filtered and residue was washed thoroughly with hot distilled water, rinsed once with 10% HCl and twice with ethanol and finally 3 times with petroleum ether. It was allowed to dry and scrapped

12

into the crucible and dried over night at 105°C in an air oven. It was then removed and cooled in a desiccator. The sample was weighed (W_1) and ashed at 600°C for 90 minutes in a muffled furnace. It was finally cooled in a dessicator and weighed again (W_2). The percentage crude fibre was calculated using equation.

$$\% \text{ Crude fibre} = \frac{W_1 - W_2}{W_0} \times 100 \%$$

Carbohydrate Content Determination

The method of AOAC (1990) was adopted where the total proportion of carbohydrate in the sample was calculated by subtracting the % sum of food nutrients: % protein, % crude fibre and % ash from 100%, as

% CHO = 100% - (% crude protein + % crude fibre + % ash + crude lipid).

Table 1 Lipid content of the tiger nut rhizomes

VARIETIES	% LIPID
Yellow	25.50
Brown	26.10

Table 2 Physical properties of the tiger nut rhizomes oils

PROPERTIES	YELLOW	BROWN
Colour	Pale yellow	Yellow
Smell	pleasant	pleasant

13

Specific gravity	0.95	0.93
Refractive Index	1.510	1.488
pH	8.85	8.06
Viscosity	97226.86	89377.14
Physical state [r]	Liquid	Liquid

Table 3 Chemical properties of the oils

PROPERTIES	YELLOW	BROWN
Iodine Value	11.62	15.23
Acid Value(mg(OH)/g oil)	14.63	8.82
Peroxide Value	3.4	2.2
Saponification Value	7.91	9.82

Table 4 Proximate composition of tiger nut

	BROWN	YELLOW
Moisture content	4. 80 ±0.003	4.00 ±0.0028
Crude protein	8.07 ±0.050	7.30 ±0.038
Crude fibre	7.60 ±0.23	6.80 ±0.29

Crude fat	26.10 ±0.98	25.50 ±088
Total ash	2.03 ±0.0054	2.00 ±0.0062
Carbohydrate	51.40 ± 0.32	54.40 ±0.11

DISCUSSION OF RESULT

From table 1, the lipid content of yellow tiger nut is (25.50 %) while brown tiger nut has (26.1 %). The value of oil yield from yellow tiger nut makes it potential oil crops. The value obtained for the yellow tiger nut is very close to that recorded by (Olaofe et al., 2006), work on Bulina cotton (Bombacopsisglabra). This high value indicate that these nuts may be suggested as potential oil crop for the production of biodiesel (Dianel et al., 2000) and are used in the treatment of debility (Chevalier ,1996).

From the table 2, the physical properties of the nut showed that the oils are found to be liquid at room temperature; also all have a characteristic pleasant odour. Refractive index of the oils within the range shows the degree of purity of the oil. The oils of the yellow and brown show a very close high degree of clarity (Mosquera, et al., 1996).

The pH values of the oils are alkaline. The result obtained in this research is different from what obtained by previous researchers like, (Mosquera, et al., 1996), who reported pH of 6.72 – 6.86 for horchata de chufa. This pH is lower compared to the pH of melon seed milk (6.25), cowpea milk (6.79) and soy milk 6.6 reported by (Akubor, 1998; Nnam, 2003 and Onweluzo and Owo, 2005) respectively.

From table 3, the chemical properties of the tiger nut showed that, the iodine value of the oil in the yellow nut is lower compared to the iodine value of the brown nut. The high iodine value indicates the presence of high amount of unsaturated fatty acids. The degree of unsaturation indicated by each sample confers their characteristic liquid state at room temperature. The greater the iodine values, the higher the oil or fat to become rancid by oxidation which implies that the oil of the brown nut becomes more rancid by oxidation.

15

The peroxide value for yellow tiger nut oil is quite high compare to the brown oils which show that the oil is most susceptible to oxidative rancidity. This could be as a result of phospholipids which are associated with fats. Degumming is the method used in removing them by hydration and separation by centrifugation of the phosphotides (ISEO 2002).

The low saponification value observed for the oils could be due to the low amount of the potassium hydroxide required to effectively saponifying the oil during manufacture of soap, hence if and only if one of these oils is to be used in the manufacture of soap, the oil from the mixture is advisable to be used in the soap making (ISEO 2002).

From table 4, the result of proximate analysis of tiger nut showed that the percentage moisture content brown tiger nut is greater what obtained in yellow tiger nut. This agreed with previous worker like (Oladele and Aina, 2007), but the value of the percentage moisture of (Oladele and Aina, 2007) is lower compared to what obtained in this research. The percentage moisture obtained by (Ogbona *et al.*, 2013) is higher than what obtained by many researchers. The crude protein obtained in this research showed that brown tiger nut has 8.70 % while yellow tiger nut has 7.30 %, the higher percentage of crude protein in brown tiger nut agreed with previous researcher like (Oladele and Aina, 2007), but the value is different from the result of (Oladele and Aina, 2007) which obtained 7.15 % and 9.70 % for yellow and brown tiger nut respectively, and (Belewu and Belewu, 2007) obtained 8.07 %. While (Ogbona *et al.*, 2013) obtained 8.19 % for crude protein. It can be noticed that the percentage crude protein of this research paper and that of the previous workers is very close.

The value of the crude fibre of this research is not agreed with (Oladele and Aina 2007), in which the percentage of crude fibre in brown tiger nut is lower than yellow tiger nut. The value obtained in this research work agreed with previous workers like (Oladele and Aina, 2007 and Ogbona 2013) where their values obtained for the percentage crude fibre are 6.26 % and 5.62 % for brown and yellow and 7.50 % respectively. The percentage fat or crude fat present in brown tiger nut is greater than yellow tiger nut, this agrees with (Oladele and Aina, 2007). The percentage crude fat obtained in the research work is 26.10 % and 25.50 % respectively for both brown and yellow tiger nut respectively, this is different from what obtained in (Oladele and Aina, 2007) where percentage crude fat is 35.43 % and 32.13 % respectively. But the result obtained is the same as what obtained in (Ogbona *et al.*, 2013) 25.50%. Though majority of the authors were

16

reported that the fat content of the tiger nut varies between 22.80 % - 32.80 % (Mokady and Dolev, 1970, Addy and Eteshola, 1984, Lissen *et al.,* 1988, Temple *et al.,* 1989 and Coskuner *et al.,* 2002).

The percentage carbohydrate in this research work showed that the brown tiger nut contained lower percentage carbohydrate than yellow tiger nut, this agrees with previous researcher like (Oladele and Aina, 2007). The percentage carbohydrate obtained in this work is 51.40 % and 54.40 % for brown and yellow tiger nut respectively. This value is higher than what obtained in (Oladele and Aina, 2007) where 46.99 % and 41.22 % was obtained for yellow and brown respectively. The value was lower compared to what obtained by (Ogbona *et al.,* 2013) 58.01 %. Carbohydrate of tiger nut is composed, mainly, of starch and dietary fibre. Tiger nut contain almost twice the quantity of starch as potato or sweet potato (Coskuner *et al.,* 2002).

The total ash of the brown tiger nut is higher than yellow tiger nut in this research work. This agreed with (Oladele and Aina, 2007), but the value obtained was lower compared to what obtained by (Oladele and Aina, 2007) and closed to what obtained by (Ogbona *et al.,* 2013)

REFERENCE

Hajirostamloo, B. 2009. Comparison of Nutritional and Chemical Parameters of Soymilk and Cow milk. *World Academy of Science, Engineering and Technology.* **57:** 436-439.

Adejuyitan J.A. 2011. Tigernut processing: its food uses and health benefits. *American Journal of Food Technology.* **6**(3): 197-201. Doi: 10.3923/ajft.2011.197.201.

Belewu M.A. & Belewu K.Y. 2007. Comparative physicochemical evaluation of tiger-nut, soybean and coconut milk sources. *International Journal of Agriculture and Biology.* **9**(5): 785-787. Doi: 1560 8530/2007/09-5-785-787.

Anon. 2012. Tigernut – a local Nigerian snack.http://naturalnigerian.com/2012/07/tiger-nuts-local-nigerian-snack/ (Accessed 21 June 2012).

Bamishaiye E.I. & Bamishaiye O.M. 2011. Tiger nut: as a plant, its derivatives and benefits. *African Journal of Food, Agriculture, Nutrition and Development.* **11**(5): 5157-5170

Arnau, J.V. 2009. Milk of tiger nut milk. http://www.tigernuts.com/news_tigernut_milk_or_cow_milk.html (Accessed 2 May 2012).

Chevallier A. 1996. The encyclopedia of medicinal plants. Dorling Kindersley Press, London.

Belewu, M.A. and Abodunrin, O.A. (2006). Preparation of kunnu from unexploited rich food source: tigernut (*Cyperus esculentus*). *World Journal of Dairy and Food Sciences.* **1:** 19-21.

Anon.2005. Horchata/tigernuts milk profile. http://www.tigernuts.com/pdf/tigernuts.milk.profile.pdf (Accessed 12 April 2012).

Anon. (2012). Nutritional differences between soy- and cow's milk. http://goaskalice.columbia.edu/nutritional-differences-between-soyand-cows-milk (Accessed 12 April 2012).

Chevallier. A .(1996). The Encyclopedia of Medicinal Plants Dorling Kindersley. *London.* ISBN 9-780751-303148.

Chopra, L., Nayor. R. N S. and Chop, I.C (1986). Glossary of Indian Medicinal Plant (including the supplement). *Council of Scientific and Industrial Research*, New Delhi.

Daniel, Z and Maria, H (2000). Domestication of plants in the old world, third edition, *Oxford University Press,*198

Facciola, S (1990). Cornucopia- A source book of edible plants. *Kampong publication* ISBN 0-96280870.9

Hedrick. U.P (1972). Edible plants of the world *Dover publications* ISBN 0-486-20459-6

Institute of Shortening and Edible Oils.(2002) (ISEO)

Komarov. V. L (1968). Flora of the USSR. Israel program for scientific translation.

Rosengarten, Jnr.F .(1984). The book of edible nuts. *Walker and Co.* ISBN 0802707699.

Olaofe, O; Akintayo, E.T; Adeyeye, E.I and Adubiaro H.O (2006). Proximate composition and functional proportion of bulina cotton (Bombcapsis glasbra) seeds *Egypt Journal of Food Science* (34) 81-90.

Addy, E. O and Eteshola, E (1984). Nutritive value of a mixture of tiger nut tuber and baobab seeds. *Journal of Science Food Agriculture.* 35 : 437 – 440

Coskunner, Y., Ercan, R., Karababa, E and Nazlican, A. N (2002). Physical and chemical properties of Chufa tuber grown in the Cukurova region of Turkey. *Journal of Science Food Agriculture* 82 : 625 – 631.

Linssen, J. P. H., Cozijnsen, J. L and Pilnik, W (1989). Chufa a new source of dietary fibre. *Journal of Science Food Agriculture* 49 : 291- 296.

Monkady, S. H and Dolev, A (1970). Nutritional evaluation of tubers of cyperus esculentus. *Journal of Science Food Agriculture* 21: 211 – 214

Temple, V. J., Ojebe, T. O and Kapu, M.M (1989). Chemical analysis of tiger nut. *Journal of Science Food Agriculture* 49 : 261 – 262.

Ogbonna, A. C., Abuajah, C. I and Utuk, R. A (2013). Tiger nut a nutritious underutilized food ingrident. *Journal Food Biology.* 2(2):14 – 17

Oladele, A. K and Aina J. O (2007). Chemical composition and functional properties of flour produced from two variety of tiger nut. *African Journal Biotechnology,* 6:2473 – 2476

Morris, B.J. (1999). The chemical Analysis of food and food products: 3rd ed. India: *CBS Publisher and Distributors.*

Association of Official Analytical Chemists (AOAC),.Official Methods of Analysis. 16th Edition, *Washington*, DC, 1990, 1: 600-792.

Mosqueral , L. A., Sims, C. A., Bates, R. P and O'keefel, S. F (1996). Flavours and stability of Horchatade Chufa. *Journal of Food Science*, 61 (4) 856- 861

19

Akubo, P. I., (1998). Physico- chemical sensory characteristic of melon seed milk. *Journal of Food Science technology*, 35 : 93 – 95.

Nnam, N. M.(2003). Nutrient composition and acceptability of vegetable milk made from oil seed *.Journal of Home Economics Research*, 5: 57 – 61.

Onweluzo,J. C and Owo, O.S (2005). Stabilization potential of water soluble polysaccharides from two legumes (Deturrummissocarpum and Mukurafia gellipes) in vegetable milk; effect on selected quality characteristic. *Journal of Home Economics Research*, 6: 39-44.